~A BINGO BOOK~

Multiplication Bingo Book

COMPLETE BINGO GAME IN A BOOK

Written By Rebecca Stark

Educational Books 'n' Bingo

TITLE: Multiplication Bingo
AUTHOR: Rebecca Stark

ISBN 978-0-87386-432-9

Educational Books 'n' Bingo

Printed in the U.S.A.

MULTIPLICATION BINGO DIRECTIONS

INCLUDED:

List of Terms

Templates for Additional Terms and Clues

2 Clues per Term

30 Unique Bingo Cards

Markers

1. **Either cut apart the book or make copies of ALL the sheets. You might want to make an extra copy of the clue sheets to use for introduction and review. Keep the sheets in an envelope for easy reuse.**

2. Cut apart the call cards with terms and clues.

3. Pass out one bingo card per student. There are enough for a class of 30.

4. Pass out markers. You may cut apart the markers included in this book or use any other small items of your choice.

5. Decide whether or not you will require the entire card to be filled. Requiring the entire card to be filled provides a better review. However, if you have a short time to fill, you may prefer to have them do the just the border or some other format. Tell the class before you begin what is required.

6. There are 50 terms. Read the list before you begin. If there are any terms that have not been covered in class, you may want to read to the students the term and clues before you begin.

7. There is a blank space in the middle of each card. You can instruct the students to use it as a free space or you can write in answers to cover terms not included. Of course, in this case you would create your own clues. (Templates provided.)

8. Shuffle the cards and place them in a pile. Two or three clues are provided for each term. If you plan to play the game with the same group more than once, you might want to choose a different clue for each game. If not, you may choose to use more than one clue.

9. Be sure to keep the cards you have used for the present game in a separate pile. When a student calls, "Bingo," he or she will have to verify that the correct answers are on his or her card AND that the markers were placed in response to the proper questions. Pull out the cards that are on the student's card keeping them in the order they were used in the game. Read each clue as it was given and ask the student to identify the correct answer from his or her card.

10. If the student has the correct answers on the card AND has shown that they were marked in response to the *correct questions,* then that student is the winner and the game is over. If the student does not have the correct answers on the card OR he or she marked the answers in response to *the wrong questions,* then the game continues until there is a proper winner.

11. If you want to play again, reshuffle the cards and begin again.

Have fun!

TERMS/ANSWERS

=	30
x	32
1	35
2	36
3	40
4	42
5	44
6	45
7	48
8	49
9	54
10	56
11	60
12	63
13	64
14	72
15	81
16	90
18	DIGIT
20	FACTOR
21	MULTIPLICATION
24	ONES PLACE
25	PRODUCT
27	TENS PLACE
28	ZERO

Additional Terms

Choose as many multiplication terms as you would like and write them in the squares. Repeat each as desired.
Cut out the squares and randomly distribute them to the class.
Instruct the students to place their square on the center space of their card.

Multiplication Bingo

Clues for Additional Terms

Write three clues for each of your additional terms.

_____ 1. 2. 3.	_____ 1. 2. 3.
_____ 1. 2. 3.	_____ 1. 2. 3.
_____ 1. 2. 3.	_____ 1. 2. 3.

+ − X ÷	+ − X ÷	+ − X ÷	+ − X ÷	+ − X ÷
+ − X ÷	+ − X ÷	+ − X ÷	+ − X ÷	+ − X ÷
+ − X ÷	+ − X ÷	+ − X ÷	+ − X ÷	+ − X ÷
+ − X ÷	+ − X ÷	+ − X ÷	+ − X ÷	+ − X ÷
+ − X ÷	+ − X ÷	+ − X ÷	+ − X ÷	+ − X ÷
+ − X ÷	+ − X ÷	+ − X ÷	+ − X ÷	+ − X ÷
+ − X ÷	+ − X ÷	+ − X ÷	+ − X ÷	+ − X ÷

=	**x**
1. This sign means "equals." 2. Whatever is left of this sign has the same value as what is to the right of this sign. 3. Fill in the sign: 9 x 2 ___ 18.	1. This sign means "multiplied by." 2. Fill in the sign: 5 ___ 2 = 10. 3. The process of multiplication is denoted by this sign.
1 1. 8 x ___ = 8 2. 1 x 1 = ___ 3. 100 x ___ = 100	**2** 1. 2 x 1 = ___ 2. 1 x 2 = ___ 3. 3 x ___ = 6
3 1. 9 x ___ = 27 2. 7 x ___ = 21 3. 6 x ___ = 18	**4** 1. 2 x 2 = ___ 2. 3 x ___ = 12 3. 6 x ___ = 24
5 1. 5 x ___ = 25 2. 6 x ___ = 30 3. 8 x ___ = 40	**6** 1. 2 x 3 = ___ 2. 6 x ___ = 36 3. 7 x ___ = 42
7 1. 12 x ___ = 84 2. 8 x ___ = 56 3. 7 x ___ = 49	**8** 1. 2 x 4 = ___ 2. 12 x ___ = 96 3. 3 x ___ = 24

Multiplication Bingo

9 1. 3 x 3 = ___ 2. 8 x ___ = 72 3. 7 x ___ = 63	**10** 1. 5 x 2 = ___ 2. 5 x ___ = 50 3. 7 x ___ = 70
11 1. 11 x 1 = ___ 2. 3 x ___ = 33 3. 2 x ___ = 22	**12** 1. 2 x 6 = ___ 2. 5 x ___ = 60 3. 6 x ___ = 72
13 1. 13 x 1 = ___ 2. 3 x ___ = 39 3. 2 x ___ = 26	**14** 1. 7 x 2 = ___ 2. 2 x 7 = ___ 3. 14 x 1 = ___
15 1. 5 x 3 = ___ 2. 3 x 5 = ___ 3. 2 x ___ = 30	**16** 1. 8 x 2 = ___ 2. 4 x 4 = ___ 3. 2 x ___ = 32
18 1. 9 x 2 = ___ 2. 6 x 3 = ___ 3. 3 x 6 = ___ Multiplication Bingo	**20** 1. 5 x 4 = ___ 2. 10 x 2 = ___ 3. 4 x 5 = ___

21 1. 7 x 3 = ___ 2. 3 x 7 = ___ 3. 21 x 1 = ___	**24** 1. 8 x 3 = ___ 2. 4 x 6 = ___ 3. 12 x 2 = ___
25 1. 5 x 5 = ___ 2. ___ x 2 = 50 3. 25 x 1 = ___	**27** 1. 9 x 3 = ___ 2. 3 x 9 = ___ 3. 27 x 1 = ___
28 1. 7 x 4 = ___ 2. 14 x 2 = ___ 3. 4 x 7 = ___	**30** 1. 10 x 3 = ___ 2. 6 x 5 = ___ 3. 15 x 2 = ___
32 1. 8 x 4 = ___ 2. 16 x 2 = ___ 3. 4 x 8 = ___	**35** 1. 5 x 7 = ___ 2. 7 x 5 = ___ 3. 35 x 1 = ___
36 1. 6 x 6 = ___ 2. 18 x 2 = ___ 3. 36 x 1 = ___	**40** 1. 8 x 5 = ___ 2. 5 x 8 = ___ 3. 10 x 4 = ___

Multiplication Bingo

42	**44**
1. $7 \times 6 =$ ___	1. $11 \times 4 =$ ___
2. $6 \times 7 =$ ___	2. $4 \times 11 =$ ___
3. $42 \times 1 =$ ___	3. $22 \times 2 =$ ___

45	**48**
1. $9 \times 5 =$ ___	1. $8 \times 6 =$ ___
2. $5 \times 9 =$ ___	2. $12 \times 4 =$ ___
3. $15 \times 3 =$ ___	3. $6 \times 8 =$ ___

49	**54**
1. $7 \times 7 =$ ___	1. $9 \times 6 =$ ___
2. $1 \times 49 =$ ___	2. $6 \times 9 =$ ___
3. $49 \times 1 =$ ___	3. $27 \times 2 =$ ___

56	**60**
1. $8 \times 7 =$ ___	1. $5 \times 12 =$ ___
2. $7 \times 8 =$ ___	2. $12 \times 5 =$ ___
3. $56 \times 1 =$ ___	3. $10 \times 6 =$ ___

63	**64**
1. $9 \times 7 =$ ___	1. $8 \times 8 =$ ___
2. $7 \times 9 =$ ___	2. $16 \times 4 =$ ___
3. $63 \times 1 =$ ___	3. $64 \times 1 =$ ___

Multiplication Bingo

72	81
1. 9 x 8 = ___	1. 9 x 9 = ___
2. 6 x 12 = ___	2. 81 x 1 = ___
3. 12 x 6 = ___	3. 1 x 81 = ___

90	**Digit**
1. 9 x 10 = ___	1. Any of the numerals 1 to 9 and 0.
2. 10 x 9 = ___	2. 75 is a 2-___ number.
3. 90 x 1 = ___	3. 433 is a 3-___ number.

Factor	**Multiplication**
1. One of 2 or more expressions that are multiplied together to get a product.	1. A basic arithmetic operation. The others are addition, subtraction & division.
2. In 5 x 7 = 35, the number 5 is one; so is the number 7.	2. A process of adding an integer to itself a specified number of times.
3. In 9 x 8 = 72, the number 9 is one; so is the number 8.	3. Its inverse (opposite) operation is division.

Ones Place	**Product**
1. The place just to the left of the decimal point.	1. The result of multiplying two numbers together.
2. In the number 17, the 7 is in this place.	2. In 5 x 7 = 35, the number 35 is the ___.
3. In the number 215, the 5 is in this place.	3. In 9 x 8 = 72, the number 72 is the ___.

Tens Place	**Zero**
1. The place two to the left of the decimal point.	1. It means "none" and is neither positive nor negative.
2. In the number 849, the 4 is in this place.	2. Any number multiplied by this number is zero.
3. In the number 572, the 7 is in this place.	3. 12 x ___ = 0

Multiplication Bingo

Multiplication Bingo

30	24	Multiplication	13	9
6	x	Zero	49	28
Factor	63		36	54
Product	64	18	44	35
90	10	7	15	20

Multiplication Bingo: Card No. 1

Multiplication Bingo

Product	Factor	40	56	Multiplication
35	49	2	64	45
72	10		8	18
14	21	63	16	28
20	Zero	7	6	15

Multiplication Bingo

Product	18	49	44	Factor
10	x	3	24	27
64	Zero		45	=
63	72	90	14	40
15	6	7	16	Multiplication

Multiplication Bingo

63	45	Multiplication	6	81
42	1	24	56	Factor
36	14		9	13
18	Digit	Zero	7	2
25	20	48	15	54

Multiplication Bingo: Card No. 4

© Barbara M. Peller

Multiplication Bingo

20	9	64	2	6
42	18	3	8	x
Multiplication	54		32	11
28	81	30	16	25
49	7	Factor	63	36

Multiplication Bingo: Card No. 5

Multiplication Bingo

=	45	40	Multiplication	54
44	64	25	24	Factor
56	4		1	8
7	90	16	81	48
35	18	30	36	Digit

Multiplication Bingo: Card No. 6

© Barbara M. Peller

Multiplication Bingo

30	45	11	32	49
35	Multiplication	10	x	42
40	13		8	1
63	14	3	Product	72
7	6	16	48	=

Multiplication Bingo: Card No. 7

© Barbara M. Peller

Multiplication Bingo

36	45	5	44	1
42	Multiplication	56	54	2
Digit	60		48	9
15	63	Product	25	14
Zero	7	81	64	35

Multiplication Bingo: Card No. 8

Multiplication Bingo

8	49	10	Digit	6
25	Multiplication	36	64	45
27	30		x	5
4	20	90	32	11
14	16	3	Product	9

Multiplication Bingo

Product	44	1	56	Digit
54	2	24	x	Multiplication
60	45		13	72
90	28	25	16	27
3	35	40	20	36

Multiplication Bingo: Card No. 10

Multiplication Bingo

=	45	64	25	35
5	27	32	8	24
42	Multiplication		40	10
3	Factor	16	6	Product
4	7	30	48	49

Multiplication Bingo: Card No. 11

Multiplication Bingo

49	9	27	44	8
10	Zero	48	81	x
30	11		54	56
7	14	Multiplication	Product	42
45	5	60	4	2

Multiplication Bingo: Card No. 12

Multiplication Bingo

4	9	=	27	54
Multiplication	5	81	8	72
44	2		10	11
36	16	1	60	Product
7	28	81	30	32

Multiplication Bingo

6	Multiplication	64	8	4
2	30	27	x	45
25	13		40	3
28	16	60	1	=
7	56	72	35	36

Multiplication Bingo: Card No. 14

Multiplication Bingo

32	8	64	49	44
=	40	24	Multiplication	25
54	30		Factor	45
7	27	5	16	4
35	14	48	Digit	10

Multiplication Bingo: Card No. 15

Multiplication Bingo

1	27	5	Digit	21
56	72	11	42	13
4	9		54	10
63	2	7	32	Product
25	Tens Place	81	14	45

Multiplication Bingo: Card No. 16

© Barbara M. Peller

Multiplication Bingo

3	Ones Place	12	27	6
32	25	16	13	11
8	36		Tens Place	5
20	35	Product	64	72
90	4	49	44	9

Multiplication Bingo: Card No. 17

Multiplication Bingo

Digit	60	2	25	56
45	3	90	54	4
8	72		12	Multiplication
20	24	16	Product	40
Tens Place	27	64	Ones Place	=

Multiplication Bingo: Card No. 18

Multiplication Bingo

54	=	27	5	60
32	44	45	49	13
Ones Place	6		x	Factor
40	Tens Place	90	14	12
Multiplication	21	35	36	81

Multiplication Bingo: Card No. 19

Multiplication Bingo

60	Ones Place	44	27	x
2	10	42	90	56
9	11		63	24
20	Zero	15	14	Tens Place
18	36	21	Product	12

Multiplication Bingo: Card No. 20

Multiplication Bingo

32	=	42	27	28
9	12	1	5	30
72	35		Ones Place	64
90	49	Tens Place	20	36
63	21	48	3	14

Multiplication Bingo: Card No. 21

© Barbara M. Peller

Multiplication Bingo

Digit	40	12	Multiplication	4
56	44	Factor	5	x
2	13		30	11
Tens Place	20	14	24	6
21	3	Ones Place	72	42

Multiplication Bingo: Card No. 22

Multiplication Bingo

1	Ones Place	49	Multiplication	81
=	60	35	32	24
40	4		15	30
72	21	Tens Place	3	14
28	Zero	36	90	12

Multiplication Bingo: Card No. 23

Multiplication Bingo

1	60	6	Ones Place	5
12	48	42	56	30
11	Digit		4	72
28	15	Tens Place	3	9
18	63	21	44	Zero

Multiplication Bingo: Card No. 24

© Barbara M. Peller

Multiplication Bingo

63	42	Ones Place	64	12
24	28	32	1	x
9	5		15	Tens Place
Factor	20	Zero	21	13
81	6	2	25	18

Multiplication Bingo: Card No. 25

© Barbara M. Peller

Multiplication Bingo

12	Ones Place	40	56	Digit
90	44	5	60	1
28	15		13	63
3	Multiplication	20	21	Tens Place
11	25	64	Zero	18

Multiplication Bingo: Card No. 26

Multiplication Bingo

40	2	Ones Place	60	10
28	15	32	Tens Place	x
16	Zero		21	63
Digit	=	42	18	24
4	13	12	Factor	11

Multiplication Bingo: Card No. 27

Multiplication Bingo

54	60	Factor	Ones Place	1
10	12	15	56	13
Zero	72		11	90
Product	Digit	35	21	Tens Place
Multiplication	8	4	18	28

Multiplication Bingo: Card No. 28

Multiplication Bingo

12	60	Digit	32	8
28	90	42	11	Factor
9	15		x	Ones Place
10	20	Multiplication	21	Tens Place
1	5	18	=	Zero

Multiplication Bingo: Card No. 29

© Barbara M. Peller

Multiplication Bingo

6	Ones Place	56	8	Tens Place
24	60	40	13	x
28	4		11	42
18	=	Multiplication	21	15
20	49	Zero	12	Factor

Multiplication Bingo: Card No. 30